Mohamad N Azra
Mhd Ikhwanuddin

Portunid Crab Aquaculture

Mohamad N Azra
Mhd Ikhwanuddin

Portunid Crab Aquaculture

Reproductive Biology and Broodstock Development

Noor Publishing

Imprint
Any brand names and product names mentioned in this book are subject to trademark, brand or patent protection and are trademarks or registered trademarks of their respective holders. The use of brand names, product names, common names, trade names, product descriptions etc. even without a particular marking in this work is in no way to be construed to mean that such names may be regarded as unrestricted in respect of trademark and brand protection legislation and could thus be used by anyone.

Cover image: www.ingimage.com

Publisher:
Noor Publishing
is a trademark of
Dodo Books Indian Ocean Ltd., member of the OmniScriptum S.R.L Publishing group
str. A.Russo 15, of. 61, Chisinau-2068, Republic of Moldova Europe
Printed at: see last page
ISBN: 978-620-2-79177-9

1. INTRODUCTION

Portunid crabs are one of the main aquaculture candidate for the future seafood commercial production. However, climate changes, disease, unrestricted capture, combined with habitat destruction and pollution, low survival and inevitably puts much stress on portunid crab populations (Azra et al. 2020; Abol-Munafi and Azra, 2018; Azra et al. 2018; Abol-Munafi et al., 2017, 2016; Kader et al., 2017; Musa et al., 2017; Farouk et al., 2016; Ikhwanuddin and Abol-Munafi, 2016; Taufik et al., 2016; Azra and Ikhwanuddin, 2015; Cik-Syahrizawati et al., 2015; Ikhwanuddin et al., 2013; Talpur et al., 2013; Azra et al., 2012; Nadiah et al., 2012; Redzuari et al., 2012). Initially, farming of portunid crab (still common practice in most countries) involves capture of wild juveniles and growing them out (a process termed as 'fattening') before selling them off.

In addition, juveniles are also captured in large volume to induce molting for the production of soft-shelled crab, a delicacy that fetches a much higher price than normal hard-shelled crabs. Furthermore, almost all current hatchery operations obtain wild-berried females for their stocks. This negatively impacts the wild local population and offers no control over the parent's heritage. These factors threaten the sustainability of the portunid crab populations in the wild and eventually lead to population reduction. Thus, focus has been shifted to aquaculture production of portunid crab in order to counter the sustainability issues of their wild populations. The hatchery seed production and nursery culture of portunid crab are still in the experimental stages as high mortality due to cannibalism is often observed when stocking density is increased (Ikhwanuddin *et al.,* 2012a; b; Chang and Ikhwanuddin, 1999).

Mud crab, genus *Scylla* is the least studied species compared to the other portunid crab species within the Portunidae family due to its more confined distribution around the Indo-pacific region (Ikhwanuddin *et al.,* 2012c). However, this mud crab species is prevalent and dominant in South East Asia, especially in water bodies around Malaysia (Ikhwanuddin *et al.,* 2011c; Waiho *et al.,* 2015). Comparing portunids with greater geographical dominancy, not many in-depth studies have been conducted on certain mud crab species, especially *S. olivacea.* The research conducted on mud crab species in Malaysia is still in its early stages, with most research focused on size and distribution (Ikhwanuddin *et al.,* 2010a; b; 2009a; b).

Currently, most researchers focus on the gonad development of portunid crabs (Yang *et al.,* 2017; Muhd-Farouk *et al.,* 2016; Waiho *et al.,* 2017a; b) while, little else has been explored on other reproductive characteristics in detail. It is justifiable that the inclination of research towards portunid crab gonad development is that the gonad maturation is directly linked to spawning and overall fecundity (Ikhwanuddin *et al.,* 2012a), thus understanding and enhancing it is beneficial economically. However, the lack of in depth knowledge on the reproductive characteristics of portunid crab, especially reproductive biology and physiology can be a hampering factor in understanding its broodstock development in details and subsequently affects the optimum production of portunid crab aquaculture.

This book therefore, focuses primarily on portunid crab reproductive biology and broodstock development of the author previous works.

2. PORTUNID CRAB: FISHERIES AND AQUACULTURE STATUS

There are more than 6,800 species of brachyuran crabs distributed around the world, with over half of the species inhabiting marine waters. A few members of a single superfamily, the Portunoidea, dominate the marine crabs. The portunid crab is well known for their worldwide distribution, which makes them an attractive potential aquaculture species. Recent study shows there are four species of genus *Scylla* found in Malaysian coastal water (Fazhan *et al.,* 2017a; b). However, recent studies found that the existing of mud crab hybrid and non-indigenous of *S. serrata* from the wild (Fazhan *et al.,* 2017a; b). However, the present review will be focused on the genus of *Scylla* especially on the species of *S. olivacea*, *S. paramamosain* and *S. tranquebarica* and *Portunus pelagicus* from the genus of *Portunus*.

The United Nations Food and Agriculture Organization (FAO, 2017) was the source for annual crustacean production data for our target area. Figure 1 depicts the fishery landings trend for crustacean species over the last 15 years.

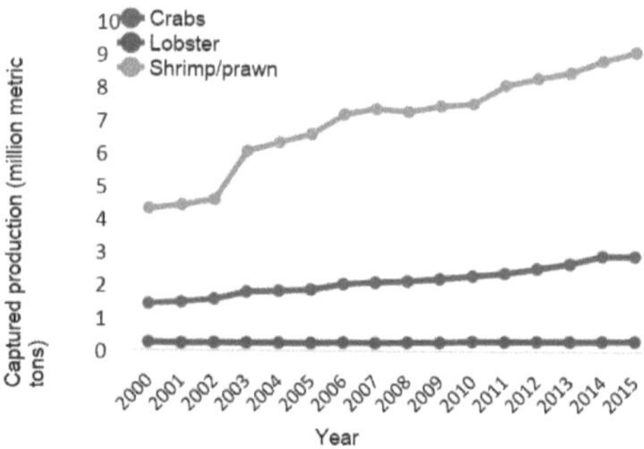

Fig. 1 Captured production of three important crustacean groups of crabs, lobster and shrimp/prawn (million metric tons)

The figure shows shrimp/prawn dominating capture production compared to crab, which has shown a slight and steady increase. The increasing trend in number of captured crabs suggests the necessity for culturing crab in order to meet production demand and recover probable decreasing total number of capture production in the future. The overall trend for capture fisheries over the last few decades is one of decline. Figure 2 shows the global fishery landings of portunid crab from 2014-2015 for the five main species *Callinectes sapidus*, *P. pelagicus*, *Charybdis* spp., *P. trituberculatus* and *S. serrata* (FAO, 2017).

Fig. 2 Trend of captured production for Portunid crabs (million metric tons)

The FAO, UN listed in their database that there are more than 340 marine fishes, crustaceans and mollusks being cultured around the world (FAO, 2017). There are 52 crustaceans with eight species of the crab, 31 of shrimp/prawn and 13 of lobster/crayfish (FAO, 2017). The probable decrease of capture fishery production in the future especially for crab has influenced its potential for being cultured. Among the crabs, portunids stand out as a high value group due to their strong export potential and high dietary value. When assessing portunid crab species as candidates for aquaculture, their life cycle, physiology and response to environmental stress need to be evaluated, as well as economic considerations.

Farming crustaceans is a rapidly growing industry that will overtake wild catching in the near future. Portunid crabs are one of the future candidates for aquaculture because of their high fecundity (Ikhwanuddin *et al.,* 2015a; 2015b; 2012a), good market price (Abol-Munafi *et al.,* 2016), high meat quality and large size (Ikhwanuddin *et al.,* 2015b; 2015c), rapid growth (Ikhwanuddin *et al.,* 2014a) and broodstock availability (Abol-Munafi *et al.,* 2017). Biologically, portunids show excellent aquaculture potential, whilst economically displaying a strong and competitive market value (Hasyima-Ismail *et al.,* 2017; Ikhwanuddin *et al.,* 2018; 2014b; 2015b).

The aquaculture industry has grown extremely fast over the past 65 years from production of less than 1 million tons in the early 1950's to 1.06 billion tons in 2015 (FAO, 2017). Regarding crustacean groups, the portunid crabs contributed 359, 041tonnes (5%) valued at USD 1.53 million (FAO, 2017). Aquaculture productions of portunid species and their value in 2015 are shown in Table 1.

Table 1 Portunid species by aquaculture production and their value from 2014 to 2015

Species	Aquaculture Production		Value (USD	
	2014	2015	2014	2015
Portunus pelagicus	41	30	197	376
Scylla olivacea	2,000	2,835	16,000	14,688
Scylla serrata	219,999	223,555	1,117,123	1,041,555
Carcinus maenas	1	3	5	27

Clearly, Giant mud crab, *S. serrata* has the highest production among the portunid crab species due to its wide distribution. Additionally, *S. olivacea* showed an impressive increase from 2014 to 2015 with a mild decrease in value (FAO, 2017) (Table 1). Traditionally, as supply increases, demand and value decreases. On the other hand, production of blue swimming crab, *P. pelagicus*, decreased from 2014 to 2015 causing their value to increase (FAO, 2017).

In Malaysia, portunid crab aquaculture is still in its preliminary phase and although small scale crab farming operations have been reported (Ikhwanuddin *et al.*, 2011a; b), the productions is considerably low and focus is primarily on crab feeding and production of soft-shelled crabs, which both rely heavily on using wild-caught juveniles. There is yet to be any fishing regulations in Malaysia that prohibit the capture of immature juvenile portunid crabs, resulting in unrestricted harvest of crabs of all sizes, especially in geographical regions with depleting wild crab resources, such as Setiu Wetlands, Terengganu.

The increasing demand for portunid crabs in Malaysia in recent years, combined with the lack of proper aquaculture production and sustainable crab fisheries have started to take a toll on the wild resources of crabs in coastal waters. Portunid crab resources are in decline. To make matter worse, water pollution and habitat destruction hastened the decline process and reduce the wild population considerably. Thus, study on the aspect of reproduction is important for sustainability of aquaculture production in the future.

3. REPRODUCTIVE BIOLOGY

3.1 Gonad Development Stages

In this sub-chapter, studies on gonadal development, size at sexual maturity for male and female, fecundity and embryonic development in portunid crabs are discussed. Reproductive biology studies on aquatic species are important to further understand their interaction with the environment, future population dynamics and life history strategies. Additionally, they provide basic information for stock management which is necessary for developing a sustainable management plan and formulating policy decisions in any fishery as spawning (fecundity) is the basis for recruitment, as well as a key factor for determining aquaculture potential of a living resource.

3.1.1 Ovarian Maturation Stages

Several studies have been carried out on the development of ovarian maturation in Portunid crabs (Ghazali *et al.*, 2017a; 2017b; 2016; Amin-Safwan *et al.*, 2016; Muhd-Farouk *et al.*, 2016; 2014; Abol-Munafi *et al.*, 2016; Ikhwanuddin *et al.*, 2015d; 2014a; 2012a). Generally, ovarian maturation stages of Portunid crab are classified into four stages. In stage 1, the whitish ovaries have a ribbon-like structure. They are located behind the hepatopancreas and their thinness makes them hard to separate. Stage 2 ovaries are light yellow and easily distinguished from the hepatopancreas and digestive glands. In stage 3, the ovaries are yellow to light orange and contain individual eggs that can be seen without magnification.

Stage 4 is deep orange to reddish orange, swollen and located behind the carapace. Stage 4 ovaries surround the hepatopancreas and contain visible eggs. Histologically, Stage 1 ovaries are filled with oogonia: primary oocytes with large nuclei and follicle cells surrounding them. During stage 2 of ovarian maturation, the size of oocytes increases with the appearance of small yolk globules in the cytoplasm of larger oocytes and continued presence of follicular cells. In stage 3, the oocytes have grown larger and contain larger yolk globules. Few follicular cells are visible. Stage 4 is characterized by massive oocytes with large yolk globules filling the entire cytoplasm so that the nucleus is barely visible. Follicular cells are almost absent.

3.1.2 Testes Maturation Stages

Recently, several studies have described the structure of the testis and discussed its functional aspects (Waiho *et al.,* 2017a; 2017b; 2016a; 2016b; Ikhwanuddin *et al.,* 2014c; Noorbaiduri *et al.,* 2014a; b). The internal reproductive system of the male portunid crab is a bilaterally symmetrical structure consisting of a pair of testes, with vas deferens and ejaculatory ducts. The highly coiled testes are grayish and opaque, and are located in the cephalothorax between the hepatopancreas and the membrane underlying the carapace. A thin, fragile commissure linking both sides of the testes is seen at the posterior end and the anterior end of the vas. Each gonad attaches to the vas deferens, which is made up of coiled tubes extending longitudinally from the end of the testis to the posterior region of the body. The vas deferens anatomically consists of three areas, the anterior (AVD), median (MVD) and posterior vas deferens (PVD).

3.2 Size at Sexual Maturity

The size at sexual maturity is one of the most important biological characteristics in the life cycle of crustaceans, including Portunid crabs and has been globally reviewed by Waiho *et al.,* (2017c). There are also various studies to determine size at sexual maturity of Portunid crabs in Malaysia coastal waters by Ikhwanuddin *et al.,* (2011a; 2009a), Waiho *et al.,* (2016b) and Muhd-Farouk *et al.,* (2017). It is considered to be a crucial determining factor of brachyurans reproductive output and the growth rate of a population as growth and maturation size influence the reproductive dynamics of a population and subsequently affect the population recruitment process in a particular geographical region. In general, sexual maturity in Portunid crabs is accompanied by a series of transformations, which eventually develop immature individuals into mature adults that are capable of reproducing and propagating.

3.2.1 Females

The sexual maturity in female Portunid crabs can be defined through morphometric analysis, ovarian maturation and the presence of spermatophores/sperm plug in spermathecae/seminal receptacle of crabs. Morphometric maturity in females includes an increase in abdomen width of the crabs. These abrupt changes occur once crabs reach maturity and have obvious functional significance where the exponential increase in abdomen width of mature females enables her to store and protect her eggs after spawning.

3.2.2 Males

The sexual maturity in male Portunid crabs can be defined through morphometric analysis, testes maturation and the presence of mating scars in male crabs. Morphometric maturity in males includes the increases of chelae size such as length, height and width. The estimation of size at morphometric maturity generally involves the use and analysis of relationships between secondary sexual characters and a reference somatic character provided that the secondary sexual characters grow at different rates during immature and mature stages. The size at maturity of male Portunid crabs should be measured using the right chela propodus length as the morphometric variable compared to carapace width.

3.3 Fecundity

Few studies have been carried out on the fecundity of portunid crabs in Malaysian waters (Ikhwanuddin *et al.,* 2014a; 2012a; 2011c). Recently, Azra *et al.,* (2015) produced a comprehensive review of fecundity studies on portunid crabs. The recent studies focus mainly on two genus of portunid crab, *Portunus* and *Scylla.* In depth knowledge of fecundity is important for management and evaluation of harvest strategies for exploited populations. In general, morphological fecundity of portunid crab consists of four stages, Stage 1 (yellowish colour), Stage 2 (orange colour), Stage 3 (brownish colour) and Stage 4 (dark grey colour). Fecundity refers to the number of eggs produced by an individual female. The number of eggs contained in the ovary of an individual crab is called the total fecundity. Fecundity can also be defined as the number of ova shed during the spawning season, the number of ripening eggs in the female prior to the next spawning season or the number of maturing eggs in the ovaries before spawning. Fecundity of Portunid crabs varies according to the species and within a species because of age, size, nutrition and water quality.

Portunid crabs show great diversity of embryonic development, especially owing to a significant variation in egg size. Females incubate their eggs, which remain attached to the pleopods, from spawning to hatching. The incubation period varies between species and other environmental factors. The estimated mean number of egg batches produced by female of Portunid crabs ranges from 1 to 6 million eggs per spawning period (Ikhwanuddin *et al.*, 2014a; 2012a; 2011c). It has been hypothesized that portunid crab fecundity is different based on the region, water body, female conditions (morphology) and ecological conditions (Ikhwanuddin *et al.*, 2014a; 2012a; 2011c). However, external morphological colour in portunid crabs does not indicate the specific stage of eggs, and thus study on their embryonic development is important for further estimation of egg stages.

3.4 Embryonic Development

Embryology is important in order to form baseline studies in assessing the quality and efficiency of eggs during hatching for a cultured species. Furthermore, such information could also ensure that prediction of hatching time would be accurate which indirectly produces higher hatching rates and enhances seed production. Scientific studies on embryonic development of portunid crabs have been conducted for the last few years in Malaysia (Ikhwanuddin *et al.*, 2016a; 2016b; 2015c; 2012a; Noor-Baiduri & Ikhwanuddin, 2015). Studies by Ikhwanuddin *et al.*, (2016a ; 2015c) indicated that there are ten (10) embryonic stages in the portunid crabs. They also recorded that several embryonic stages were found in the portunid crabs, which may vary depending on the environmental and culture conditions, breeding techniques and nutrition. Interestingly, a study by Noor-baiduri and Ikhwanuddin (2015) showed only five (5) stages of embryonic development due to the use of different breeding techniques (*in vitro* fertilization techniques) in that study.

However, the earlier classification of most crustacean species describing that there are ten (10) embryonic stages in the brachyurans is most common and is based on the amount of yolk. In previous studies, the embryonic development of portunid crabs was monitored from spawning to hatching (Ikhwanuddin *et al.*, 2016a; 2015c). Eggs were collected from the crab's clutch using a sterile surgical blade and the size of each egg was determined as the mean between its maximum and minimum diameters. The incubation period is the duration in days required by the broodstock from spawning to hatching (Azra and Ikhwanuddin, 2016). In general, the incubation period of eggs of *Scylla* spp. was between 7 to 13 days depending on the conditions maintained (Ikhwanuddin *et al.*, 2016a; 2015c). The reduced incubation periods are possibly due to adequate food supplies and good water quality parameters.

4. BROODSTOCK DEVELOPMENT

The development of methods for increasing Portunid crab broodstock targets improvement of seed production. Broodstock conditioning usually entails mature crabs being brought into the hatchery, housed in tanks and allowed to mature or ripen over time under culture conditions. By obtaining broodstock from the wild, it is possible to enhance broodstock at the hatchery. Commonly, portunid crab broodstock are brought into the hatchery and kept for a period of 2-3 weeks before spawning and maturation process. The period of conditioning depends on the reproductive condition of the crab when they enter the hatchery. Most studies done on methods to enhance broodstock are based on environmental conditions (Musa et al., 2017; Waiho et al., 2016b; Adnan et al., 2016; Amin-Safwan et al., 2016; Azra et al., 2015; Ikhwanuddin et al., 2014d; Noor-baiduri et al., 2014b), nutritional requirements (Ikhwanuddin et al., 2017; 2018; 2015d, 2015e; Kader et al., 2017; Ghazali et al., 2017a; 2017b; Azra and Ikhwanuddin, 2016; Azra et al., 2015), physiological effects (Muhd-Farouk et al., 2016; 2014), genetic selection and information (Waiho et al., 2017b; Yang et al., 2017), healthy culture conditions and parasite infestation (Cik-Syahrizawati et al., 2015; Talpur et al., 2013; Waiho et al., 2017d) and other broodstock management practices such as eyestalk fablation and limb autotomy techniques, inter-species mating in captivity (Fazhan et al., 2017c; 2017d; Farouk et al., 2016; Nadiah et al., 2012). These methods help improve Portunid crab broodstock, thus increasing hatchery seed production and ensuring a sustainable future for mass crab culture.

4.1 Environmental Conditions

Research efforts have been targeted towards the development of improved broodstock for several crustacean species including Portunid crabs. Several studies were done to assess the effects of various environmental factors on the broodstock production of brachyuran crabs (Musa *et al.*, 2017; Waiho *et al.*, 2016b; Amin-Safwan *et al.*, 2016; Adnan *et al.*, 2016; Azra *et al.*, 2015; Waiho *et al.*, 2015; Ikhwanuddin *et al.*, 2014d). The effects of different water salinities on *S. olivacea* broodstock was conducted by Amin-Safwan *et al.* (2016) by using LC_{50} assessment in nine different water salinities (2, 5, 10, 15, 20, 25, 30, 35 and 40 ppt). In another study, Ikhwanuddin *et al.* (2014d) and Noorbaiduri *et al.* (2014b) conducted experiments on the effects of salinity on the mating success of portunid crabs, especially *S. olivacea*. The crabs were reared at salinities of 15, 20, 25 and 30 ppt for 30 days and the results showed no significant relationship between salinity level and mating success of Portunid crabs, especially *S. olivacea*. However, results did show that 25 ppt salinity was optimal for mud crab pairs to achieve successful sperm deposition in mating. Adnan *et al.* (2016) also showed that salinity levels affected maturation of ovaries in Portunid crabs, but not the 17β-estradiol levels.

4.2 Nutritional Requirements

Studies of Portunid crab broodstock production revealed the importance of diet in increasing the number of berried females and the health of their offspring. Portunid crab broodstock are fed using a natural, artificial or mixed diet containing traditional low-cost items such as clams, mussels, squid, shrimp, and fish (Ikhwanuddin *et al.*, 2018; 2015d; Kader *et al.*, 2017; Ghazali *et al.*, 2017a, b; Azra & Ikhwanuddin, 2016). However, frequent use of natural foods can cause water quality to worsen, with adverse effects on growth and maturation (Azra & Ikhwanuddin, 2016). Therefore, the use of artificial products is suggested to maintain water quality and increase broodstock output. Research on formulated feed for Portunid crab broodstock is important for the development of inexpensive, nutritious commercial feeds. It is recommended that natural diets be supplemented with artificial ingredients to promote spawning, improve egg quality and larval production. Most artificial diets consist of fish, shrimp or squid meal with fish or squid oil, wheat and binder. The use of fishmeal in commercial hatchery production increases the total production cost which may lead to increases in the price of the species cultured. However, a new approach to overcome dependence on fishmeal combines various natural and artificial ingredients for improved larval quality of the cultured species. The inclusion of more protein improves the quality of the formulated diet promoting reproduction of the broodstock.

4.3 Physiological Effects

To increase seed production, hatcheries need to develop a better understanding of the influence of physiology on broodstock selection. Currently, there is little data on the negative effects of chemicals on crab biology and broodstock production (Muhd-Farouk *et al.*, 2016; 2014). Most of the compounds for Portunid crab culture were tested in the lab and are very expensive. Some of these additives such as the steroid hormone, 17-hydroxypregnenolone (Muhd-Farouk *et al.*, 2014) and vertebrate steroid hormones (Muhd-Farouk *et al.*, 2016) induced ovarian maturation and accelerated reproduction of *S. olivacea*, However, they cannot be used commercially until they are proven safe. Muhd-Farouk *et al.* (2016; 2014) concluded that 17α-hydroxypregnenolone at a dose of 0.01 μg BW^{-1} dose produced the largest number of mature ovary stages, stimulated ovary maturation with formation of large oocytes and a high gonad somatic index.

4.4 Genetic Selection and Information

The field of crab aquaculture needs to be updated by increased research to identify the genes involved in gonadal development with the goal of using this data to enhance reproduction and broodstock. More details are needed in the areas of genetic selection for improved reproductive performance and methodology for easily obtaining the genealogy of Portunid crab species. Identifying the genes that are differentially expressed under various physiological stages can provide the essential information for developing better methods of crab culture. Recent studies using quantitative reverse transcription polymerase chain reaction (qRT-PCR) have identified the sex gene in the Portunid crab, *S. paramamosain*, Yang *et al.* (2017) as well as a comprehensive analysis of the transcriptome data derived from testis tissue of *S. olivacea* (Waiho *et al.*, 2017b) at different maturation stages using the Illumina HiSeq. Yang *et al.*, (2017) found 316 differentially expressed genes by comparing the transcriptomes at different stages. Waiho *et al.* (2017b) identified 156,181 single-nucleotide polymorphisms by sequence matching from the testes at different maturation stages. For example, the beta crystallin-like gene proved to be highly up-regulated at immature stages and down-regulated during maturation.

4.5 Healthy Culture Conditions and Parasite Infestation

Producing disease free and healthy crab broodstock is an essential first step in order to develop successful hatchery production capability for a culture species. Previous studies indicated that microbial infections (Cik-Syahrizawati *et al.*, 2015; Talpur *et al.*, 2013; Talpur & Ikhwanuddin, 2012a), parasite infestation (Waiho *et al.*, 2017d) and overexploitation, threatened the global fishery industry especially the aquaculture of brachyuran crabs (Portunidae). These risks can seriously hinder aquaculture activities for crabs by limiting the healthy broodstock available for hatchery seed production. In brief, Talpur *et al.* (2013) showed that more than five species of pathogen were found in the gut of portunid crabs from the genus *Vibrio*, *Pseudoalteromonas*, *Staphylococcus* and *Micrococcus*. The study by Waiho *et al.* (2017d; 2018) found that parasitization of crabs with rhizocephalan barnacles, *Sacculina beauforti*, resulted in abnormal size, weight, abdomen width and gonopod or pleopod length.

4.6 Other Broodstock Management Practices

Several methods to enhance portunid crab reproduction have focused on the induction of the gonad, a key organ for embryonic development and egg production. A few studies have described the importance of the gonad in synthesizing vetellogenin under controlled conditions (Fazhan *et al.*, 2017c; 2017d; Farouk *et al.*, 2016; Nadiah *et al.*, 2012). A method that is easier and known to improve and accelerate reproduction in portunid crabs is eyestalk ablation. Removing the mud crab's eyestalk changes its hormonal regulation, which is vital for regulating its growth, metamorphosis, metabolism, reproduction, development and behavior. Many crustacean species in culture are incapable of natural maturation, which places a high demand on wild caught populations. Therefore, to promote broodstock maturation in captivity, a study by Ikhwanuddin *et al.* (2018) was carried out to determine the effect of eyestalk ablation on the ovarian maturation stages of the orange mud crab, *S. olivacea*. The highest percentage of ovarian Stage 4 maturation was found in the ablated group (12.5%) compared to the control group (0%) and ovarian development in ablated animals was significantly different. The study concluded that eyestalk ablation was the best technique for increasing ovary maturation. Nadiah *et al.* (2012) found that limb autotomy could trigger successful mating, vitellogenin production and ovary maturation in Portunids.

5. CONCLUSION AND RECOMMENDATION

Crab farming is not a new invention. Wild-caught crabs have been collected and raised in ponds for hundreds of year. However, modern aquaculture of Portunid crabs has only recently taken center-stage as the preferred commercial method. This review attempts to showcase the current fundamental knowledge on reproductive biology, broodstock management, breeding activities and larval rearing of portunid crabs that are paving the way for sustainable aquaculture development in the future. The fundamental knowledge on Portunid crab aquaculture has been acquired by researchers in the Asia-Pacific region including Malaysia allowing for notable rapid progress in culture methods over the past decade.

Numerous enquiries from the general public, researchers and farmers about portunid crab aquaculture over the past couple years has inspired the need for an academic book that describes the fundamental knowledge of portunid crab reproduction, the foundation for the production of high-quality crablets for sustainable aquaculture production. More advanced studies on seed production techniques are required to further enhance the crablet quality of Portunid crabs. This requires further research programs on selective breeding and quality broodstock development to be carried out in collaboration with local industry and international research institutions. Collaborations such as these are promising and expected to contribute greatly to the domestication, farming and larval rearing of portunid crabs, both regionally and locally.

REFERENCES

Abol-Munafi, A.B., Azra, M.N. 2018. Climate change and the crab aquaculture industry: problems and challenges. *J. Sustain. Sci. Manag.* 13, 1-4.

Abol-Munafi, A.B., Pilus, N., Amin, R.M., Azra, M.N., Ikhwanuddin, M. 2017. Digestive enzyme profiles from foregut contents of blue swimming crab, *Portunus pelagicus* from Straits of Johor, Malaysia. *J. Assoc. Arab Univ. Basic Appl.* Sci. 24, 120-125.

Abol-Munafi, A.B., Mukrim, M.S., Amin, R.M., Azra, M.N., Azmie, G., Ikhwanuddin, M. 2016. Histological profile and fatty acid composition in hepatopancreas of blue swimming crab, *Portunus pelagicus* (Linnaeus, 1758) at different ovarian maturation stages. *Turk. J. Fish. Aquat. Sc.* 16, 251-258.

Adnan, A.S., Musa, N., Ikhwanuddin, M. 2016. *Effects of water salinity on female Scylla olivacea in captivity: morphological, histological characteristics of ovary and the levels of reproductive steroid hormone 17-beta-estradiol.* 106 pp. Lambert, Academic Publishing. Saarbrücken, Germany.

Amin-Safwan, A., Muhd-Farouk, H., Nadirah, M., Ikhwanuddin, M. 2016. Effect of water salinity on the external morphology of ovarian maturation stages of orange mud crab, *Scylla olivacea* (Herbst, 1796) in captivity. *Pak. J. Biol. Sci.* 19, 219-226.

Azra, M.N., Ikhwanuddin, M. 2016. A review of maturation diets for mud crab genus *Scylla* broodstock: Present research, problems and future perspective. *Saudi J. Biol. Sci.* 23, 257-267.

Azra, M.N., Ikhwanuddin, M. 2015. Larval culture and rearing techniques of commercially important crab, *Portunus pelagicus* (Linnaeus, 1758): Present status and future prospects. *Songklanakarin J.Sci. Technol.* 37, 135-145.

Azra, M.N., Wendy, W., Talpur, A.D., Abol-Munafi, A.B., Ikhwanuddin, M. 2012. Effect of tank colouration on survival, growth and development rate of blue swimming crab, *Portunus pelagicus* (Linnaeus, 1758) larvae. *Int. J. Curr. Res. Rev.* 4, 117-123.

Azra, M.N., Abol-Munafi, A.B., Ikhwanuddin, M. 2015. A Review of broodstock improvement to brachyuran crab: reproductive performance. *Int. J. Aquac.* 5, 1-10.

Azra, M.N., Chen, J.C., Ikhwanuddin, M., Abol-Munafi, A.B. 2018. Thermal tolerance and locomotor activity of blue swimmer crab *Portunus pelagicus* instar reared at different temperatures. *J. Therm. Biol.* 74, 234-240.

Azra, M.N., Aaqillah-Amr, M.A., Ikhwanuddin, M., Ma, H., Waiho, K., Ostrensky, A., Tavares, C.P.D.S. and Abol-Munafi, A. B. 2020. Effects of climate-induced water temperature changes on the life history of brachyuran crabs. *Rev Aquacult*, 12: 1211-1216.

Chang, W.W.S., Ikhwanuddin, M. 1999. Pen culture of mud crabs, genus *Scylla*, in the mangrove ecosystems of East Sarawak, Malaysia, pp. 83-88. *In*: Keenan, C. P. and A. Blackshaw (eds) *Mud crab Aquaculture biology*. Proceedings of an international scientific forum held in Darwin, Australia, 21-24 April 1997, ACIAR Proceedings No. 78, 216 p.

Cik-Syahrizawati, M.Z., Ikhwanuddin, M., Wendy, W., Zulhisyam, A.K., Lee, S.W. 2015. Antibiotic and heavy metal resistance of bacteria isolated from diseased mud crab (*Scylla serrata*). *J. Trop. Resour. Sustain. Sci.* 3, 1-5.

FAO, Food and Agriculture Organization (FAO). 2017. Dataset of Global Capture Production. United Nations (UN), Rome, Italy.

Farouk, H. M., Ikhwanuddin, M., Jasmani, S. 2016. *Inducing the ovarian maturation of orange mud crab, Scylla olivacea.* 58 pp. Lambert, Academic Publishing. Saarbrücken, Germany.

Fazhan, H., Waiho, K., Ikhwanuddin, M. 2017a. Non-indigenous giant mud crab, *Scylla serrata* (Forskål, 1775) (Crustacea: Brachyura: Portunidae) in Malaysian coastal waters: a call for caution. *Mar. Biodivers. Rec.* 10, 26.

Fazhan, H., Waiho, K., Azri, M.F.D., Al-Hafiz, I., Wan Norfaizza, W.I., Megat, F.H., Jasmani, S., Ma, H., Ikhwanuddin, M. 2017b. Sympatric occurrence and population dynamics of *Scylla* spp. in equatorial climate: Effects of rainfall, temperature and lunar phase. *Estuar. Coast. Shelf S.* 198: 299-310

Fazhan, H., Waiho, K., Wan Norfaizza, W.I., Megat, F.H., Ikhwanuddin, M. 2017c. Inter-species mating among mud crab genus *Scylla* in captivity. *Aquaculture.* 471, 49-54.

Fazhan, H., Waiho, K., Wan Norfaizza, W.I., Megat, F.H., Ikhwanuddin, M. 2017d. Assortative mating by size in three species of mud crabs, genus *Scylla* (Brachyura: Portunidae). *J. Crustac. Biol.* 37, 654-660.

Ghazali, A., Azra, M.N., Noordin, N.M., Abol-Munafi, A.B., Ikhwanuddin, M. 2017a. Ovarian morphological development and fatty acids profile of mud crab (*Scylla olivacea*) fed with various diets. *Aquaculture.* 468, 45-52.

Ghazali, A., Noordin, N.M., Abol-Munafi, A.B., Azra, M.N., Ikhwanuddin, M. 2017b. Ovarian maturation stages of wild and captive mud crab, *Scylla olivacea* fed with two diets. *Sains Malays.* 46, 2273-2280.

Ghazali, A., Ikhwanuddin, M., Bolong, AMA. 2016. *Ovarian maturation stages of orange mud crab, Scylla olivacea: morphological, histological and fatty acids characteristics of ovary.* 108 pp. Lambert, Academic Publishing. Saarbrücken, Germany

Hasyima-Ismail, N., Amin-Safwan, A., Fairuz-Fozi, N., Megat, F.H., Muhd Farouk, H., Kamaruddin, S.A., Ikhwanuddin, M., Ambak, M.A.

2017. Study on carapace width growth band counts relationship of orange mud crab, *Scylla olivacea* (Herbst, 1796) from Terengganu coastal waters, Malaysia. *Pak. J. Biol. Sci.* 20, 140-146.

Ikhwanuddin, M., Ghazali, A., Nahar, S.F., Wendy, W., Azra, M.N., Abol-Munafi, A.B. 2018. Testes maturation stages in mud crab, *Scylla olivacea* broodstock fed different diets. *Sains Malaysi.* 47, 427-432.

Ikhwanuddin, M., Azmi, W.A., Borkhanuddin, M.H., Rahmah, S., Idris, I. 2017. Improving the health of Setiu Wetlands ecosystems and productivity of crustacean resources for livelihood enhancement. *J. Sustain. Sci. Manag.* SI3: i-v.

Ikhwanuddin, M., Abol-Munafi, A.B. 2016. Fish and shellfish domestication and stock enhancement: Current status and future directions. *Asian J. Sci. Res.* 9, 167-170.

Ikhwanuddin, M., Azra, M.N., Noorulhudha, N.F., Siti-Aishah, A., Abol-Munafi, A.B. 2016a. Embryonic development and hatching rate of blue swimming crab, *Portunus pelagicus* (Linnaeus, 1758) under different water salinities. *Turk. J. Fish. Aquat. Sc.* 16, 669-677.

Ikhwanuddin, M., Hayimad, T., Ghazali, A., Halim S.S.A., Abdullah, S.A. 2016b. Resistance test on early larval stage of blue swimming crab, *Portunus pelagicus*. *Songklanakarin J.Sci. Technol.* 38, 83-90

Ikhwanuddin, M., Khairil, I.O., Azra, M.N., Waiho, K. 2015a. Biological features of sentinel crab *Podophthalmus vigil* (Fabricus, 1798) in Terengganu coastal water, Malaysia. *J. Fish. Aquat. Sc.* 10, 501-511.

Ikhwanuddin, M., Yin, T.P., Menon, A.J., Jasmani, S., Bolong A.M.A. 2015b. Effect of different cryoprotectants and sperm densities of orange mud crab, *Scylla olivacea* (Herbst, 1796) for long-term storage of spermatozoa. *Sains Malays.* 44, 1283–1288.

Ikhwanuddin, M., Lan, S.S., Abdul-Hamid, N., Fatihah-Zakaria, S.N., Azra, M.N., Abdullah, S.A.,Ambok-Bolong, A.M. 2015c. The embryonic development of orange mud crab, *Scylla olivacea* (Herbst, 1796) held in captivity. *Iran.J. Fish. Sci.* 14, 885-895

Ikhwanuddin, M., Mohamed, S., Rahim, A.I.A., Azra, M.N., Jaaman, S.A., Bolong, A.M.A. and Noordin, N.M. 2015d. Observations on the effect of natural diets on ovarian re-maturation in blue swimming crab *Portunus pelagicus* (Linnaeus, 1758). *Indian J. Fish.* 62, 124-127.

Ikhwanuddin, M., Azmie, G., Helmi, M.G., Waiho, K., Fazhan, H., Bachok, Z. 2015e. Indigenous occurrences of green mud crab, *Scylla paramamosain* (Estampador, 1949) (Crustacea: Decapoda: Portunidae) from Setiu Wetlands, Peninsular Malaysia. *In*: Mohamad, F., Salim, J.M., Jani, J.M. and Shahrudin, R. (Eds). *Setiu Wetlands: Species, Ecosystem and Livelihood.* 228 pp. Penerbit Universiti Malaysia Terengganu (UMT)

Ikhwanuddin, M., Jamal, N.A., Abol-Munafi, A.B. and Muhammad-Farouk, H. 2014a. Reproductive biology on the gonad of female orange mud crab, *Scylla olivacea* from the west coastal water of peninsular Malaysia. *Asian J. Cell Biol.* 9, 14-22

Ikhwanuddin, M.,Liyana, A.N., Azra, M.N., Bachok, Z., Abol-Munafi, A.B. 2014b. Natural diet of blue swimming crab, *Portunus pelagicus* at Strait of Tebrau, Johor, Malaysia. *Sains Malays.* 43, 37-44

Ikhwanuddin, M., Muhd-Farouk, H., Memon, A.J., Wendy, W., Abol-Munafi, A.B. 2014c. Sperm viability assessment over elapsing time maintained at 2°C of orange mud crab, *Scylla olivacea* (Herbst, 1796). *Pak.J. Biol. Sci.* 17, 1069-1073

Ikhwanuddin, M., Noor-Baiduri, S., Wan-Norfaizza, W.I., Abol-Munafi, A.B. (2014d). Effect of water salinity on mating success of orange mud crab, *Scylla olivacea* (Herbst, 1796) in captivity. *J. Fish. Aquat. Sci.* 9, 134-140.

Ikhwanuddin, M., Azra, M.N., Sung, Y.Y., Abol-Munafi, A.B., Long, M.S. 2013. Growth and survival of blue swimming crab (*Portunus pelagicus*) reared on frozen and artificial foods. *Agr. Sci.* 4, 76-82

Ikhwanuddin, M., Azra, M.N., Siti-Aimuni, H., Abol-Munafi, A.B. 2012a. Fecundity, embryonic and ovarian development of blue swimming crab, *Portunus pelagicus* (Linnaeus, 1758) in coastal water of Johor, Malaysia. *Pak. J. Biol. Sci.* 15, 720-728

Ikhwanuddin, M., Talpur, A.D., Azra, M.N., Mohd Azlie, B., Hii, Y.S., Abol-Munafi, A.B. 2012b. Effects of stocking density on the survival, growth and development rate of early stages blue swimming crab, *Portunus pelagicus* (Linnaeus, 1758) larvae. *World App. Sci. J.* 18, 379-384

Ikhwanuddin, M., Azmie, G., Juariah, H.M., Abol-Munafi, A.B., Zakaria, M.Z., Ambak, M.A. 2012c. Tracking the movement of mud crabs, Genus *Scylla* from mangrove area using telemetry system. *Borneo Sci.* 30, 40-56.

Ikhwanuddin, M., Azmie, G., Juariah, H.M., Zakaria, M.Z., Ambak. M.A. 2011a. Biological information and population features of mud crab, genus *Scylla* from mangrove areas of Sarawak, Malaysia. *Fish. Res.* 108, 299-306.

Ikhwanuddin, M., Azmie, G., Zakariah, M.I., Abol-Munafi A.B. 2011b. *Mud Crab: Culture System and Practices in Malaysia.* Penerbit Universiti Malaysia Terengganu (UMT)

Ikhwanuddin, M., Muhamad, J.H., Long S.M and Abol-Munafi Ambok Bolong. 2011c. Fecundity of blue swimming crab, *Portunus pelagicus* Linnaeus, 1758 from Sematan fishing district, Sarawak coastal water of South China Sea. *Borneo J. Resour. Sci. Technol.* 1: 46-51.

Ikhwanuddin, M., Bachok Z., Mohd-Faizal W.W.Y., Azmie, G., Abol-Munafi, A.B. 2010a. Size of maturity of mud crab *Scylla olivacea* (Herbst, 1796) from mangrove areas of Terengganu coastal waters. *J. Sustain. Sci. Manag.* 5, 134-147

Ikhwanuddin, M., Bachok Z., Hilmi M.G., Zakaria M.Z. 2010b. Species diversity, carapace width-body weight relationship, size distribution and sex ratio of mud crab, genus *Scylla* from Setiu Wetlands of Terengganu coastal waters, Malaysia. *J. Sustain. Sci. Manag.* 5, 97-109.

Ikhwanuddin, M., Shabdin, M.L., Abol-Munafi, A.B. 2009a. Size at maturity of blue swimming crab (*Portunus pelagicus*) found in Sarawak coastal water. *J. Sustain. Sci. Manag.* 4, 56-65.

Ikhwanuddin, M., Abol-Munafi, A.B., Shabdin, M.L. 2009b. Catch information of blue swimming crab (*Portunus pelagicus*) from Sarawak coastal water of South China Sea *J. Sustain. Sci. Manag.* 4, 93-103.

Kader, M.A., Bulbul, M., Asaduzzaman, M., Abol Munafi, A.B., Noordin, N.M., Ikhwanuddin, M., Ambak, M.A., Ghaffar, M.A., Ali, M.E. 2017. Effect of phospholipid supplements to fishmeal replacements on growth performance, feed utilization and fatty acid composition of mud crab, *Scylla paramamosain* (Estampador 1949). *J. Sustain. Sci. Manag.* SI3: 47-61.

Muhd-Farouk, H., Amin-Safwan, A. Arif, M.S., Ikhwanuddin, M. 2017. Biological information and size at maturity of male crenate swimming crab, *Thalamita crenata* from Setiu Wetlands, Terengganu coastal waters. *J. Sustain. Sci. Manag.* 12, 119-127

Muhd-Farouk, H., Jasmani, S., Ikhwanuddin, M. 2016. Effect of vertebrate steroid hormones on the ovarian maturation stages of orange mud crab, *Scylla olivacea* (Herbst, 1796). *Aquaculture* 451, 78-86

Muhd-Farouk, H., Abol-Munafi, A.B., Jasmani, S., Ikhwanuddin, M. 2014. Effect of steroid hormones 17α-hydroxyprogesterone and 17α-hydroxypregnenolone on ovary external morphology of orange mud crab, *Scylla olivacea. Asian J. Cell Biol.* 9, 23-28

Musa, N., Manaf, M.T.A., Saari, N.A., Hamzah, N., Ibrahim, W.N.W., Aznan, A.S., Zakaria, K., Ghani, S.H.A., Razzak, L.A., Musa, N., Zainathan, S.C., Wahid, M.E.A., Shaharom-Harrison, F., Ambak, M.A., Ikhwanuddin, M.,, Ghaffar, M.A. 2017. Some aspects of population biology of edible orange mud crab, *Scylla olivacea* (Herbst, 1796) during pre and post monsoon in Setiu Wetland, Terengganu, Malaysia. *J. Sustain. Sci. Manag.* SI3: 29-37.

Nadiah, W.N., Ikhwanuddin, M.,Abol-Munafi, A.B. 2012. Remarks on the mating behavior and success of blue swimming crab, *Portunus pelagicus* (Linnaeus, 1758) through the induction of limb autotomy technique. *J. Anim. Vet. Adv.* 11, 1149-1157.

Noorbaiduri, S., Ikhwanuddin, M. 2015. Artificial crablets production of orange mud crab, *Scylla olivacea* (Herbst, 1796) through *in vitro* fertilization technique. *J. Fish. Aquat. Sci.* 10, 102-110.

Noorbaiduri, S., Abol-Munafi, A.B., Ikhwanuddin, M. 2014a. Acrosome reaction stage of sperm from mud crab, *Scylla olivacea* (Herbst, 1796): mating in wild and in captivity. *J. Fish. Aquat. Sci.* 9, 237-244.

Noor-Baiduri, S., Nurul-Akmal, S., Ikhwanuddin, M. 2014b. Mating success of hybrid trials between two mud crab species, *Scylla tranquebarica* and *Scylla olivacea*. *J. Fish. Aquat. Sci.* 9, 85-91.

Redzuari, A., Azra, M.N., Abol-Munafi, A.B., Aizam, Z.A., Hii, Y.S., Ikhwanuddin, M. 2012. Effects of feeding regimes on survival, development and growth of blue swimming crab, *Portunus pelagicus* (Linnaeus, 1758) larvae. *World App. Sci. J.* 18, 472-478

Talpur, A.D., Memon, A.J., Khan, M.I., Ikhwanuddin, M., Danish Daniel, M.M., Abol-Munafi, A.B. 2013. Gut Lactobacillus sp. bacteria as probiotics for *Portunus pelagicus* lariculture: effects on survival, water quality and digestive enzyme activities. *Invertebr. Reprod.Dev.* 57, 173-184.

Taufik, M., Bachok, Z., Azra, M.N., Ikhwanuddin, M. 2016. Effects of various microalgae on fatty acid composition and survival rate of the blue swimming crab *Portunus pelagicus* larvae. *Indian J. GeoMarine Sci.* 45, 1512-1521.

Waiho, K., Fazhan, H., Jasmani, S., Ikhwanuddin, M. 2017a. Gonadal development in males of the orange mud crab, *Scylla olivacea* (Herbst, 1796) (Decapoda, Brachyura, Portunidae). *Crustaceana* 90, 1-19

Waiho, K., Fazhan, H., Shahreza, M.S., Moh, J.H.Z., Noorbaiduri, S., Wong, L.L., Sinnasamy, S., Ikhwanuddin, M. 2017b. Transcriptome analysis and differential gene expression on the testis of orange mud crab, *Scylla olivacea*, during sexual maturation. *PlosOne* 12, e0171095

Waiho, K., Fazhan, H., Baylon, J.C., Madihah, H., Noor-Baiduri, S., Ma, H., Ikhwanuddin, M. 2017c. On types of sexual maturity in brachyurans, with special reference to size at the onset of sexual maturity. *J. Shellfish Res.* 36, 807- 839

Waiho, K., Fazhan, H., Glenner, H., Ikhwanuddin, M. 2017d. Infestation of parasitic rhizocephalan barnacles *Sacculina beauforti* (Cirripedia, Rhizocephala) in edible mud crab, *Scylla olivacea*. *PeerJ* 6, 3419

Waiho, K., Fazhan, H., Baylon, J.C., Norfaizza, W.I.W., Ikhwanuddin, M. 2016a. Use of abdomen looseness as an indicator of sexual maturity in male mud crab *Scylla* spp. *J. Shellfish Res.* 35, 1027-1035

Waiho, K., Fazhan, H., Ikhwanuddin, M. 2016b. Size distribution, length–weight relationship and size at the onset of sexual maturity of the orange mud crab, *Scylla olivacea*, in Malaysian waters. *Mar. Biol. Res.* 12, 726-738

Waiho, K., Mustaqim, M., Fazhan, H., Norfaizza, W.I.W., Megat, F.H., Ikhwanuddin, M. 2015. Mating behaviour of the orange mud crab, *Scylla olivacea*: The effect of sex ratio and stocking density on mating success. *Aquacult. Rep.* 2, 50-57

Yang, X., Ikhwanuddin, M., Li, X., Lin, F., Wu, Q., Zhang, Y., You, C., Liu, W., Cheng, Y., Shi, X., Wang, S., Ma, H. 2017. Comparative transcriptome analysis provides insights into differentially expressed genes and long non-coding RNAs between ovary and testis of the mud crab (*Scylla paramamosain*). *Mar. Biotechnol.* 20, 20-34.

ACKNOWLEDGEMENTS

This works is supported by the Ministry of Higher Education under the Higher Center of Excellence (HiCOE) grant for development of future food through breeding technology, under the Sustainable Shellfish Aquaculture. The first author would like to appreciate the Ministry of Higher Education, Malaysia for the opportunity given to pursue a Postdoctoral programme in the Institute of Tropical Aquaculture and Fisheries, Universiti Malaysia Terengganu.

Related Figures with Legends

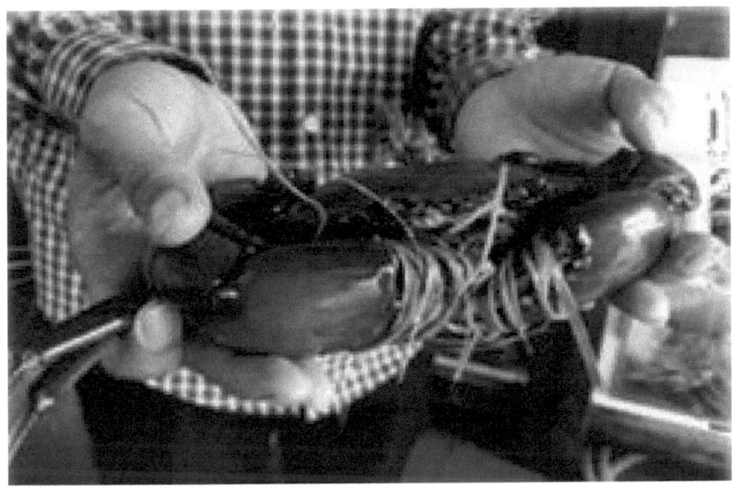

Mud crab, genus *Scylla* one of the most wanted crabs in Malaysia

Berried females of mud crabs, genus *Scylla* with early stage of embryonic development

Berried females of mud crabs, genus *Scylla* with middle stage of embryonic development

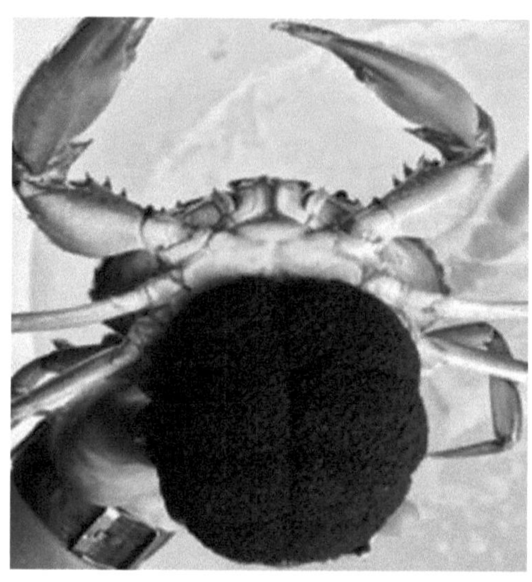

Berried females of mud crabs, genus *Scylla* with later stage of embryonic development

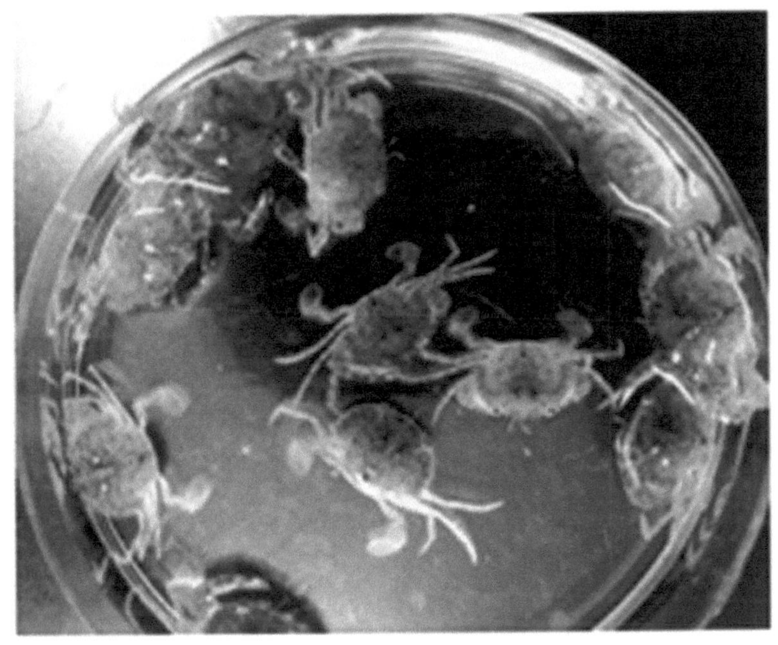

Crablet production of mud crab, genus *Scylla*

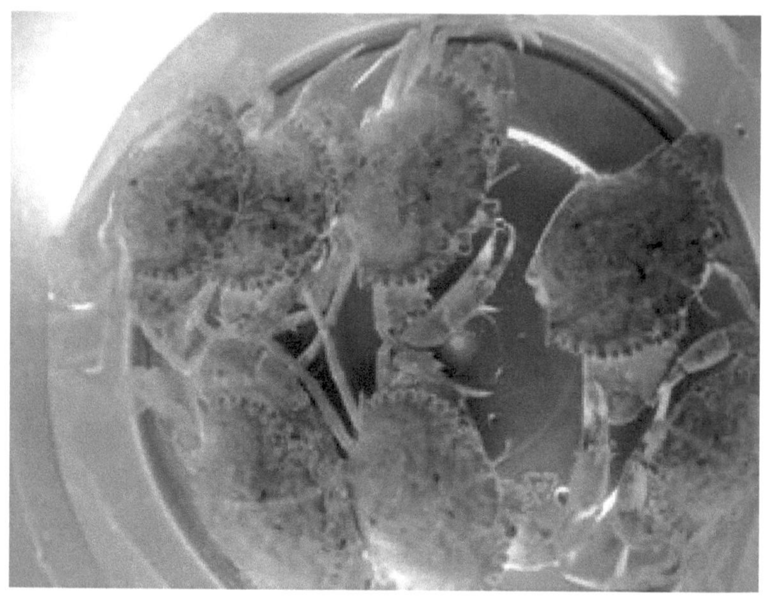

Juvenile of mud crab, genus *Scylla* produced in hatchery of Institute of Tropical Aquaculture and Fisheries, Universiti Malaysia Terengganu

Soft-shell mud crab, potential candidate for downstream research in the future

Formulated feed for the commercial broodstock development in captivity

Mud crabs, genus *Scylla* after harvested

Male mud crab, genus *Scylla* based on the abdominal shape

Matured stage of female mud crab, genus *Scylla*

Immature stage of female mud crab, genus *Scylla*

Juvenile crabs produced in the hatchery was being released in the crab ponds

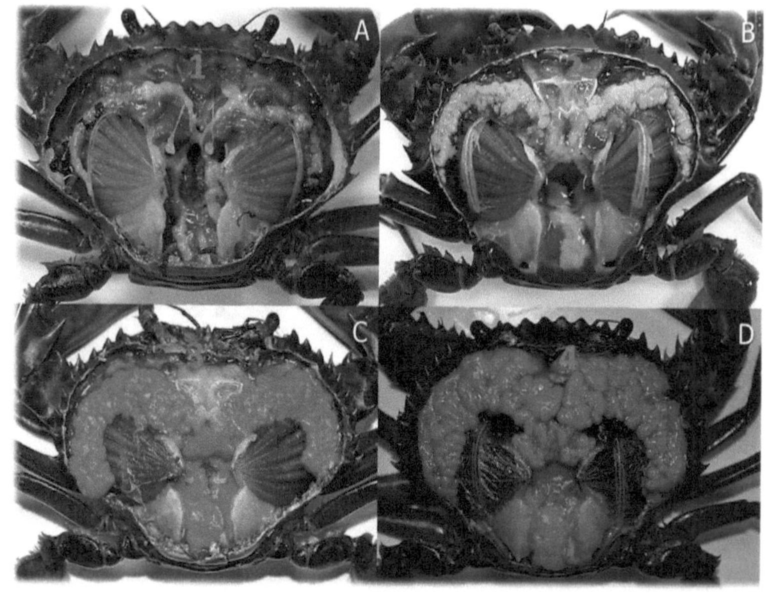

Four stage of ovarian maturation in mud crab, genus *Scylla*

Testis maturation of mud crab, genus *Scylla*

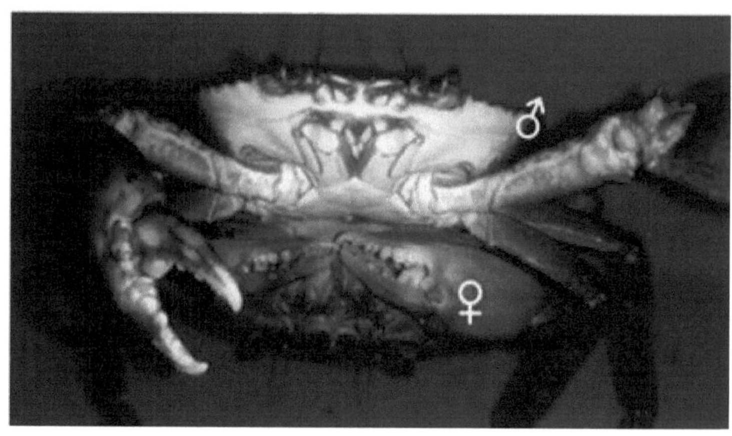

Mating activity in mud crab, genus *Scylla* in captivity

An example of small crab pond in East Coast of Peninsular Malaysia

I want morebooks!

Buy your books fast and straightforward online - at one of world's fastest growing online book stores! Environmentally sound due to Print-on-Demand technologies.

Buy your books online at
www.morebooks.shop

Kaufen Sie Ihre Bücher schnell und unkompliziert online – auf einer der am schnellsten wachsenden Buchhandelsplattformen weltweit! Dank Print-On-Demand umwelt- und ressourcenschonend produziert.

Bücher schneller online kaufen
www.morebooks.shop

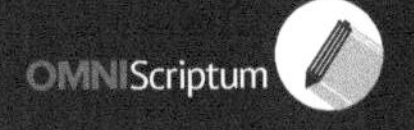

KS OmniScriptum Publishing
Brivibas gatve 197
LV-1039 Riga, Latvia
Telefax: +371 686 204 55

info@omniscriptum.com
www.omniscriptum.com

OMNIScriptum